YOUR KNOWLEDGE HAS VALUE

- We will publish your bachelor's and master's thesis, essays and papers

- Your own eBook and book - sold worldwide in all relevant shops

- Earn money with each sale

Upload your text at www.GRIN.com and publish for free

Bibliographic information published by the German National Library:

The German National Library lists this publication in the National Bibliography;
detailed bibliographic data are available on the Internet at http://dnb.dnb.de .

Imprint:

Copyright © 2008 GRIN Verlag, Open Publishing GmbH
Print and binding: Books on Demand GmbH, Norderstedt Germany
ISBN: 9783640445554

This book at GRIN:

http://www.grin.com/en/e-book/137307/process-optimisation-of-injection-moulding-
using-computer-simulation

Kyaw Lin

Process Optimisation of Injection Moulding using Computer Simulation

GRIN Publishing

University of South Australia

PROCESS OPTIMISATION OF INJECTION MOULDING USING COMPUTER SIMULATION

Research Proposal

Submitted by: Kyaw Lin

Abstract

Determining optimal values of process control factors is critical in injection moulding process because of their influences on product quality, productivity and cost of production. In the past, many researchers exploited traditional and artificial methods but the limitations of different methods prevented to achieve optimal level of process design variables for multiple-input multiple-output (MIMO) injection moulding process. To bridge the gap, this study aims to develop a computer integrated optimisation system (CIOS). In this research, virtual reality (VR) technology is employed, in the first phase, for simulation purpose and combination of design of experiments (DOE) and SAPSO-based ANN method is used for optimising the simulation results in order to achieve global optimal solution for the control parameters. Apparently, the proposed approach is a new integration system that can help the users determine optimal parameter settings to accomplish the MIMO injection moulding process with competitive benefits of cost and production efficiency.

Table of contents

Figures

Tables

Symbols

Xa1, Xa2 = variables of injection velocity level 1 and 2

Xb1, Xb2 = variables of melt temperature level 1 and 2

Xc1, Xc2 = variables of mould temperature level 1 and 2

Xd1, Xd2 = variables of packing pressure level 1 and 2

Xe1, Xe2 = variables of gate location level 1 and 2

E1, E2 = error term of level 1 and 2

k = the number of factors

n_l = the experiment level

m = the number of replicates

n_g = the number of groups (levels),

n_j = the number of individuals in the jth group,

x_{ij} = the ith individual in the jth group,

$\bar{x} =$ the mean of the jth group,

$\bar{x}' =$ the grand mean value.

$y_i =$ expected output.

$o_{ij} =$ the predicted output of the network.

$no =$ number of output nodes.

$np =$ number of training set samples.

$w =$ inertia weight.

$v =$ velocity of the particle.

$c_1, c_2 =$ acceleration constants.

$r_1, r_2 =$ random numbers in the range $[0, 1]$.

$x_i(t) =$ the current position of a particular particle.

$p_{id} =$ the best one of the solution that the current particle has reached.

$p_{gd} =$ the best one of the solutions that all particles have reached.

$prob =$ pre $-$ determined probability .

$temp =$ temperature.

$Itemp =$ initial temperature (constant).

Acronyms

MIMO - multiple-input multiple-output

CIOS - computer integrated optimisation system

VR - virtual reality

DOE - design of experiments

ANNs - Artificial Neural Networks

BPNN - Back Propagation Neural Network

GRNN - General Regression Neural Network

GA - Genetic Algorithm

DFP - Davidon-Fletcher-Powell

SA - Simulated Annealing

DEA - Data Envelopment Analysis

MRSN Ratio - Multi-Response Signal to Noise Ratio

ANP - Analytic Network Process

PSO - particle swarm optimisation

FDM - Finite Difference Method

FEM - Finite Element Method

FVM - Finite Volume Method

BEM - Boundary Element Method

CPU - Central Processing Unit

CVFEM - Control Volume Finite Element Method

EMM - Equivalent Medium Method

CV - Control Volume

FE - Finite Element

VIMS - Virtual Injection Moulding System

VR - Virtual Reality

FEA - Finite Element Analysis

MS - Motion Simulation

SV - Scientific Visualisation

FV - Finite Volume

RSM - Response Surface Methodology

CFD - Computational Fluid Dynamics

MPC - Model Predictive Control

AMD - absolute mean deviation

VSM - Virtual Simulation Model

CAD - Computer Aided Design

ML - Model Loader

MS - Moulding Simulator

MD - Motion Definer

COM - Coupled Optimisation Model

ANOVA - analysis of variance

UniSA - University of South Australia

LTU - Language and Teaching Unit

PROCESS OPTIMISATION OF INJECTION MOULDING USING COMPUTER SIMULATION

1. Introduction

1.1. Project background

Injection moulding is one of the most important processes for manufacturing commercial items due to its advantages such as short product cycles and easily moulded complicated shapes. It can be basically divided into four stages: plasticisation, filling, packing and cooling, and ejection (Rosato & DV Rosato, 1995). Of four stages, filling phase includes the process parameters that influence on the product quality as well as the productivity. The high demand of three dimensional parts for new application areas in different industries has promoted the development of its efficiency. Previously, trial-and-error approach based on the engineer's experience and intuition was employed to identify the significant parameter setting. However, it is costly, time-consuming and inappropriate for complex manufacturing processes (Lam, Deng & Au 2004). Manzione (ed. 1987) suggested that a key to improve the capacity of handling important aspects of injection moulding process is the utilisation of computer simulation. However, there remains a gap between simulation results and optimisation objectives owing to the complexity and dynamic interaction of the injection moulding process. Moreover, the issues regarding materials, product design and quality, production time, and cost effectiveness make it more complicated. Accordingly, the requirement of more flexible optimisation method draws attention of many researchers. The ideal method is to check and measure the effects of process parameters in the real production line, but it has many disadvantages such as time-consuming, high-cost and inaccuracy (Bikas, Pantelelis & Kanarachos 2002, p.112). Therefore, in most cases, optimisation algorithms are integrated with computer simulation to bridge the gap between simulation and reality.

1.2. Project aims and problem statement

The purpose of this research is to develop a new sophisticated computer integration system to optimise the injection moulding process using computer simulation and optimisation algorithms. In dealing with this project, the study will focus on the three sub-problems:

i. The first sub-problem is to determine the process design variables that are dominant in the injection moulding process.

ii. The second sub-problem is to decide on the optimisation algorithms that are able to quickly search the best results for the moulding process performance.

iii. The third sub-problem is to discover the most efficient computer simulation package for injection moulding.

1.3. Scope of the project

To eliminate irrelevancies to the identified objective and the research problem, the scope of the project is clearly stated as follows:

i. The study will be limited to plastic injection moulding process for most of manufacturing enterprises use plastic as a major material for both household and industrial products.

ii. The study will not attempt to optimise all stages of the injection moulding process but filling stage as it is critical.

iii. The study will be limited to the dominant control process design variables of the filling phase.

iv. The study will not evaluate the economic factors of the injection moulding process.

v. The study will not appraise simulation software used in other industries.

1.4. Hypotheses

i. The appropriate combination of process control factors influence on the production efficiency and product quality.

ii. The preferred simulation package can facilitate the optimisation purpose, giving more flexibility in understanding and decision making of the injection moulding process.

iii.   The optimisation algorithms designated are able to effectively search the global optimum solution for the entire process.

2. Literature Review and Discussions

In the literature, a large number of papers regarding optimisation of injection moulding process can be found. The word "optimisation" in different papers means making an improvement of process factors affecting the product quality and productivity within the feasible limits that satisfy all specified constraints. These factors include design variables relating with process parameter settings and process design. Besides, the use of optimisation methods and computer simulations plays an important role for optimising the plastic injection moulding process.

2.1. Review on work about process control variables

A number of researches have been carried out with different aspects of moulding quality to optimise the process parameters that comprise not only physical and chemical parameters such as melt temperature, mould temperature, injection velocity, injection pressure, and packing pressure, but also time parameters such as injection time, packing time and cooling time (Chen et al. 2007; Kurtaran & Erzurumlu 2006). In the past studies, different combination of process parameters had been used. Wu and Liang (2005) used six control process parameters: melt temperature, mould temperature, packing pressure, injection velocity, injection acceleration and packing time to examine the effects of parameters on the width of weld line of an injection moulded plastic part. Chen, Wang, Fu, & Chen (2008) employed four process control parameters: Injection Time, Velocity Pressure Switch Position, Packing Pressure, and Injection Velocity to identify the optimum parameter setting for the moulded plastic push-button housing piece under consideration of single quality response: product weight. Chiang and Chang (2006) applied four process factors: melt temperature, mould temperature, injection pressure, and injection time to determine the optimal process parameter for a thin-shell plastic part with multiple quality characters. Zhou, & Turng (2007) exploited seven process parameters: melt temperature, mould temperature, packing pressure, injection time, packing time, cooling time, and velocity/pressure switch-over (V/P) by volume to optimise the volume shrinkage for Plexiglass optical lens with different thickness at the centre and outer rim.

In terms of process design optimization, a series of studies have been found, which focus on different perceptive of the moulding process design: gate location, runner system, cooling system, and part geometry. Subramanian, Tingyu, & Seng, (2005) attempted to optimise the gate location in order to minimize the warpage of the injection moulded part, analysing the distortion of the legs and reference pads in a plastic optical housing for a CD optical pickup. Zhai, Lam, & Au (2006) tried to specify the optimum solutions of gate locations and runner size of a multi gated moulding process, considering two critical affects: weld line and warpage. Lee, & Lin (2006) targeted to determine the optimal runner and gating system parameters for a multi-cavity injection mould in order to minimise the warp formation. Deng, Zheng, & Lu (2008) aimed to optimise multi-class variables including process parameter, gating system, runner system, cooling system and part geometry under consideration of multi-response moulding qualities: part warpage, weld lines, air traps and so on. Reviewing the above papers, one can conclude that appropriate combination of process parameters and process design variables is a very important issue to achieve the objective quality requirements.

2.2. Review on previous work about process optimisation

Concerning the optimisation methods used for the injection moulding process, there are a significant number of papers in the literature, some of which focus on single criterion while another on multiple criteria. The techniques used for optimisation issue include traditional methods and artificial intelligence methods. Previously, trial-and-error method was used to determine the process parameter settings, depending on the experience and intuition of the engineers. However, it has many drawbacks and is unsuitable for complex processes to get the actual optimum results (Lam et al. 2004). Next, Taguchi's parameter design method has been used for experimental design and process improvement as a central one in many papers. Liao et al. (2004) exploited L_{27} orthogonal array experimental tests based on Taguchi's method to optimise the process conditions of a thin-wall injection moulding of a cellular phone cover made of amorphous PC/ABS resin plastics, taking into account the interaction effects between process parameters, and quality targets: shrinkage and warpage. Oktem, Erzurumlu & Uzman (2007) utilised Taguchi optimization tool to determine plastic injection moulding process parameters for thin shell parts. In this case, material property, part design, and injection-moulding conditions were considered as the factors or variables needed to be changed to minimize shrinkage. Nonetheless, Taguchi's method is only able to find the best combination of parameter level that includes discrete values so it falls short when the parameter values are continuous (Chen et al. 2008). To deal with the continuous parameter variables, Artificial Neural Networks (ANNs) has been introduced as an alternative means (Chiu et al. 1997). Chen et al. (2009) claims that ANNs can map the relationship between input factors and output responses and it has different sub-categories such as Back Propagation Neural Network (BPNN), General Regression Neural Network (GRNN). However, it finds difficult to search the final optimal variables. Later on, the robust optimisation method Genetic Algorithm (GA) has been extensively used for random searching of the global optimum value in large dimensional space without being trapped in the local optimum (Tseng 2006; Shen, Wang & Li 2007). Lam, Den & Au (2006) argues that GA is opportunistic but not deterministic alone. Therefore, many researchers combine different nature of optimisation methods to approach the optimal design and process conditions. Chen et al. (2008) integrated Taguchi's parameter design method, BPNN, and Davidon-Fletcher-Powell (DFP) method to optimize the process parameter settings of plastic injection moulding, considering single product quality (weight). Furthermore, Chen et al. (2009) combined Taguchi's parameter design method, BPNN, GA and engineering optimization concepts to optimize the process parameters, for an experiment on a standard plastic piece, under multiple-input multiple-out (MIMO) consideration. They claim that their approach can effectively help engineers determine optimal process parameter settings and achieve competitive advantages of product quality and the costs.

On the other hand, some researchers used different optimisation methods for their particular objectives. Su & Chang (2000) proposed the combination of Neural Network (NN) and Simulated Annealing (SA) to optimise the parameter design of the injection moulding process. Loera et al. (2008) used Data Envelopment Analysis (DEA) optimisation strategy for a thermoplastic injection moulding, considering multiple criteria: design and process variables to meet several performances. Upon the application of their approach, they discussed the optimum parameter settings for the moulding of rear automotive lamps to control the part dimension and surface properties (aesthetic) using seven control process inputs. Moreover, Park & Ahn (2004) used DOE to reduce the cooling time and the injection pressure. This research proved that DOE is useful to discover the cause and effect relationship between the inputs and outputs of a process, and to determine the optimal process parameters with fewer testing trials. Furthermore, Deng (2008) coupled Multi-Response Signal to Noise Ratio (MRSN Ratio) with Analytic Network Process (ANP) to optimize the process parameters of multiple-response process in order to achieve higher production efficiency. MRSN is one of reformed Taguchi's methods for multiple-quality response process parameter optimization and ANP is a systemic process that applies ratio scales to evaluate internal relationship of dimensions, criterions, and alternatives. They used five control parameters: mould temperature, pipe temperature, injection velocity, injection pressure and packing time to achieve the targeted four quality responses: weight, length (dimensional), warpage & shrinkage (surface property) for case study of a bottom cover of polypropylene Modem. In addition, Da & Xiurun (2004) presented SAPSO-based ANN method using particle swarm optimisation (PSO) with simulated annealing (SA) in order to achieve global optimum. Their approach modelled the relationship between confining pressure, peak stress and corresponding strain, and showed its flexibility in escaping local optimum. For that reason, it can be concluded that proper selection and/or combination of different techniques to meet quality criterion and production efficiency becomes one of the most important factors in optimisation of injection moulding process.

2.3. Review on optimisation techniques used for injection moulding

Towards the optimisation of injection moulding process, development of computer simulation makes it easier to analyse the process in the early design stage before real implementation. As a result, it becomes essential tool in the injection moulding industry. Quite a large number of different simulation software has been introduced in the last three decades, and a number of applications have been published. While some researchers used only computer simulation for optimisation purposes, others integrated simulation tools with optimisation algorithms.

2.3.1. Computer simulation

Basically, computer simulation is built up with sequential development of mathematical equation groups to analyse physical features of material inside the mould cavity. Kim and Turng (2004) outlined the development process of computer–aided simulation technique from traditional 2.5-D Hele-Shaw approach to 3-D simulation applications. Disadvantages and limitations of the 2.5-Dimensional solution evidently pointed the necessity of building a new method for moulding simulation. As a result, they introduced a typical procedure implemented with four numerical methods: Finite Difference Method (FDM), Finite Element

Method (FEM), Finite Volume Method (FVM) and Boundary Element Method (BEM), which are considered as key supplementary tools for fulfilling simulation objectives. FDM is relatively effective and quite simple for solving partial and differential set of algebraic equations. It is, however, difficult to apply this method to a highly complicated boundary. Hence, the application of it is restricted to regular and simple domains. Conversely, FEM has excellent flexibility in solving complex boundaries and irregular geometries. Nevertheless, the global matrix system employed by this method requires large memory space and CPU (Central Processing Unit) time to process the data. Meanwhile, BEM successfully tackles the excessive effort of computation, an inherent weak point of most simulation algorithm, but it is impaired when non-linear problems for the matrix system of algebraic equations are dense and non-symmetric. Although FVM is a particular case of FDM, it can overcome the troubles of FDM by subdividing physical domain into small control volumes. Therefore, the combination of this method and FEM can create an application of simulation algorithm to solve complicated domain and complex geometries but less CPU time and computational effort.

Alternatively, the two other methods, Control Volume Finite Element Method (CVFEM) and Equivalent Medium Method (EMM) were introduced by Dong (2005) to solve the simulation problem in a vacuum assisted moulding process. CVFEM bases on the matrix of permeability tensors and the pressure gradients of the three directions combined with Gauss's theorem and Darcy's law to generate simulation data. Similarly, EMM solution is generated by constructing a matrix of elements based on 3-D mesh generation of the model. This method requires fewer elements than that of traditional CVFEM but much complex mesh generation algorithm. To support and fulfil for the EMM method, a mesh generation algorithm is also organisationally produced and presented in the same article. EMM enhances the reduction of time consumption for computational activities with 85% time savings compared to that of other traditional method (Dong, C 2006, p.1212).

Consequently, based on the steps of a basic simulation procedure, Shojaei et al. (2003) presented two methods for tracking simulation algorithm, exploiting Quasi-Steady State and Nodal Partial Saturation approach. Quasi-Steady State approach is constructed to generate the algorithm for isothermal models while Nodal Partial Saturation approximation is for non-isothermal solutions. More than algorithms, Quasi-Steady State and Nodal Partial Saturation approach are intensively developed into higher applications of computer code with given names RTMS and RAPFIL, respectively. Both of those numerical schemes are, then, combined together and implemented in FORTRAN 77. Moreover, Quasi-Steady State and Nodal Partial Saturation algorithm are suitable for personal computer but can not reach high accuracy. There is an error up to 1.52% in filling time compared to practical experiment (Shojaei et al. 2003).

Additionally, Shojaei (2006) studied numerical simulation of filling process of moulding in full three-dimensional domain using CVFEM and developed numerical algorithm to progress the flow front based on a quasi-steady state approach. The numerical algorithm he presented was coded and the resulting computer code was used to predict the necessary parameters of the filling stage such as flow progression, pressure distribution and mould clamping force for the mould containing both the single and multi-layer performs. Shen, Wang, & Li (2007) integrated a hybrid FE/ FD method with a Control Volume (CV) to simulate the filling stage of injection molding. Also, they developed a Visual C++ program for compressible flow analysis of the filling stage, generalising the Hele–Shaw flow of a compressible viscous fluid under non-isothermal conditions with consideration of the effects of compressibility and phase change. Llado´ & Sa´nchez (2008) employed Finite Element (FE) Moldflow numerical simulation software to determine the cause of blush that appears around the gate in the injection of PVC fittings and to predict the relationship between blush and injection rate and melt temperature. Zhou, Shi, & Ma (2009) proposed a Virtual Injection Moulding System (VIMS) integrated with Virtual Reality (VR), Finite Element Analysis (FEA), Motion Simulation (MS) and Scientific Visualisation (SV). VR technology can provide the imperative view of the model and process, offering interactive (what-if) studies. Obviously, their research shows that VIMS can highlight the problematic area and influences of the design variables for the target product requirements. Conversely, application of computer simulation alone cannot solve the very complicated manufacturing cases because of its nature: only to run the program not to solve.

2.3.2. Integration of computer simulation and optimisation algorithms

Subsequently, in most cases, several researchers integrate the computer simulation with optimisation algorithms. Chang & Yang (2001) proposed a collocated implicit Finite Volume (FV) approach with the SIMPLE segregated algorithm to solve the three-dimensional injection mould filling problems. The numerical model they described dealt with three-dimensional isothermal flow of incompressible, high-viscous Newtonian fluids with moving interfaces. Turng and Peic (2002) integrated a CAE simulation tool (C-MOLD) with a process optimization program (OPTIMUS) to determine the optimal design and process variables for injection moulding. A polynomial type of Response Surface Methodology (RSM) was used in OPTIMUS to determine the relationship between the process design objectives and the settings of design and process parameters. Gerber, Dubay, & Healy (2006) combined Computational Fluid Dynamics (CFD) with Model Predictive Control (MPC) to improve and control polymer melt temperature within the barrel and nozzle region, considering a three-heater zone interaction. They used CFX-TASCflow for the CFD simulations model that acts as a slave to the MPC optimisation algorithm. Deng, Zheng & Lu (2008) integrated the CAE (Moldflow) software with Particle Swarm Optimisation (PSO) algorithm to optimise the multi-class design variables of the injection moulding such as part thickness, process parameters and gate location, intending to achieve the multi-response quality requirements: warpage, weld lines, air-traps and so on. As PSO was designed to search Pareto optimality (Loera et al. 2008), their experiment on plastic AC power outlet got the complete optimality, without normalisation of objective functions and specification of the weight factors. The different levels of successes in the above reviewed papers show the strength of integrating different techniques: computational, numerical, mathematical to approach the optimum process design variables.

2.4. Conclusion and discussions

The literature shows that optimisation of process control parameter is important to achieve product quality requirements in order to improve the efficiency of plastic injection moulding process. Although there are a number of studies that focused on the impact of process parameters, few of them focused on combination of process and design factors. Moreover, most papers used only optimisation methods in searching optimal solution. The literature suggests that few researches carried out with integration of computer simulation and optimisation methods are more efficient to attain the global optimum. Therefore, there is a need to develop this assessment that can help injection moulding industry worldwide from certain aspects of the production efficiency. For that reason, this research will effort to develop an integration system that can optimise the impacts of process control factors, and the research findings are likely to be useful for further studies on the same issue.

3. Methodology

This project is attempting to develop a computer integrated optimisation system (CIOS) for injection moulding process, targeting high product quality as well as production efficiency. The CIOS integrates computer simulation with optimisation methods using C++ programming language to achieve global optimal solution of the control parameter settings for MIMO plastic injection moulding.

3.1. Sampling control factors

In sampling the data for experiment, Stratified Sampling Design (Leedy & Ormrod 2005) will be used, setting certain criteria among the population of each group. In this study, two levels of data from each group will be used as a sample that represents all the characteristics of population of each group. To avoid bias, the sampling procedure is carefully planned for control factors as follows:

To demonstrate the effectiveness of the proposed approach, four criteria for quality responses are selected as the milestones of the project. The first two dimensional properties are length and weight, then the latter are warpage and shrinkage belonged to surface properties. The quality responses of length and weight have "nominal the best" characteristic. Length has a specification with limited tolerance whereas weight does not have a certain target value. For the time being, the quality responses of warpage and shrinkage have "the smaller the better" characteristic. After setting the objectives for optimisation purpose, process control factors are determined as the input data. From the literature review, four process factors dominant in the injection moulding process and influential to the quality responses are selected. They are injection velocity that influences the mechanical and dimensional stability of the product, melt temperature that has strong relation with the polymer viscosity, filling velocity, filling pressure, and cavity pressure-time profile, mould temperature that is critical for cycle time and warpage formation, and packing pressure that affects the product shrinkage and warpage. While proper gate location can improve the warpage and production efficiency, it is taken into account as a process design variable. The formulation of process control factors with two levels can be seen in Table.1.

Table.1.Process control factors and levels

Factor	Level 1	Level 2
A: Injection velocity (mm/sec)	X_{a1}	X_{a2}
B: Melt temperature (°C)	X_{b1}	X_{b2}
C: Mould temperature (°C)	X_{c1}	X_{c2}
D: Packing pressure (MPa)	X_{d1}	X_{d2}
E: Gate location	X_{e1}	X_{e2}
Error	E_1	E_2

3.2. Data collection and data analysis

This research will be implemented through combination of qualitative and quantitative approach in continue mode for system building and conducting experiment, respectively. Secondary data will be used due to its benefits such as easy to collect, less time-consuming for data collection, analysis and interpretation (Blaxter, Hughes & Tight 2001). The data required for system phenomenon will be collected from the literature: technical reports, recent and latest researches. As the proposed system uses new techniques, numerous form of data will be collected and examined from various angles to discover the problems existing within the phenomenon. To evaluate the effectiveness of the system and verify the validity of the theories, a case study will be followed based on practical experiment.

The data from the experiment results will be collected using statistical spreadsheet via longitudinal study. Then the sorted data will be set into tables and charts to analysis and interpret the significance of data whereas cause and effect relationship of inputs and outputs will be discovered by studying data correlation. Leedy and Ormrod's (2005) randomised two-factor design will be used for experiment design.

After collecting the data, this research will use absolute mean deviation (AMD) to compare the performance of length quality. In the meantime, sample standard deviation will be used to compare the performance of four quality responses and sample mean to compare the performance of warpage and shrinkage quality response. The comparison of the quality response results will be tabulated as shown in Appendix A.

Furthermore, data analysis will be followed up with null hypothesis test and t-test to confirm the hypotheses previously set in the project.

3.3. Framework of proposed approach

The study pays high attention to the filling stage of the plastic injection moulding process, considering multiple responses. The continuous interaction of the flow pattern, temperature and pressure profiles as well as control parameters makes this stage the most complicated phase out of four. Besides, the nature of plastic material is more viscous and compressible in molten stage so the melt (fluid) and its flow will be considered as non-Newtonian compressible fluid under non-isothermal flow, taking into account of the effects of heat and pressure. After determining the objectives and constraints, for process optimisation, a new integration system is proposed. In this approach, VR technique presented by Zhou, Shi, & Ma (2009) will be used for simulation purpose to analyse the variables and highlight the problematic area. Then, the statistical results from simulation will be imported to coupled optimisation model that includes DOE, SA and PSO. The schematic diagram of the flow of proposed approach can be seen in Figure.1.

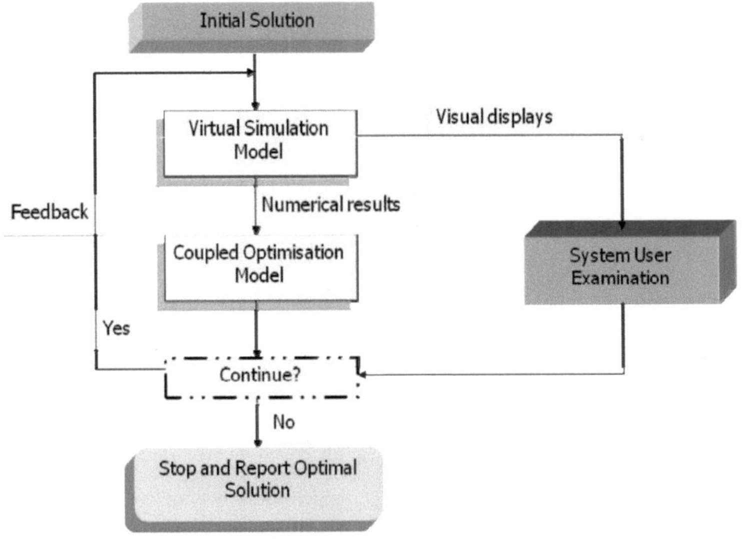

Figure.1. Flow diagram of proposed CIOS

3.4. Virtual Simulation Model (VSM)

The main purpose of employing VSM is to facilitate visualisation and optimisation of the injection moulding process based on VR technology. "VR is the use of computer generated virtual environments and the associated hardware to provide the user with the illusion of physical presence within that environment which allows the designer to 'virtually manufacture' the product while designing it" (Ma et al. 2007, p. 1093). This system includes three main parts: virtual environment, pre-processing and post-processing as shown in Figure.2.

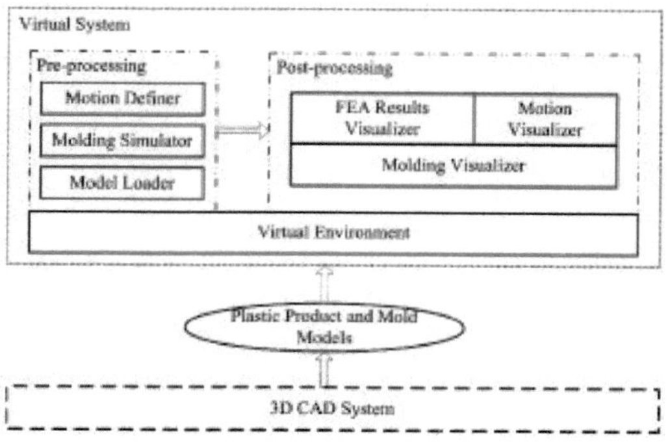

Figure.2. Structure of VSM (Zhou, Shi & Ma 2009, p. 299)

In application of this model, necessary information will be firstly supplied for creating prototypes of plastic product and mould design, which are inputs of the virtual system and will be done in commercial 3D CAD (Computer Aided Design) software. It can be seen that CAD system is used as an external task to create models of the products and mould design.

3.4.1. Virtual environment

The virtual environment is used for visualisation of mono-mode and stereo-mode of prototypes. The latter mode can provide 3D stereoscopic view by means of an emitter and a CrystalEyes Workstation that is a wireless set of liquid crystal shutter eyewear for Stereo3DTM imaging. The shutter eyewear generates a stereoscopic feeling by synchronizing with the display device, showing the images to the left eye and right eye alternatively via a switch. The former mode requires no emitter and glasses as it does not provide stereoscopic display.

3.4.2. Pre-processing

With regard to the pre-processing, preparation work will be done in each component namely Model Loader (ML), Moulding Simulator (MS) and Motion Definer (MD). ML is used to read and manage the product part and mould design, and to display them in the virtual environment. MS is applied for an interface of process control parameter setting and presentation of simulation results based on FEA and CFD. MD is for grouping all mould parts and generation of the movement pathways for moulding simulation.

3.4.3. Post-processing

In post-processing phase, FEA results visualiser is employed to validate the moulding process in terms of numerical results for flow patterns, process and design parameters, and quality related factors and so on. By analysing the results, the possible faults and inadequacies can be evidently highlighted. Meanwhile, the effects of the control parameters such as melt temperature, gate location on the mouldability and product quality can be studied by changing them interactively. The motion visualiser is used to display the mould motion and the machine operations. The moulding visualiser is to combine FEA results, moulding machine motion and mould assembly so the moulding cycle can be viewed intuitively.

3.4.4. Application of VSM

Fundamental steps of the application of VSM are demonstrated as follows. The process starts with creation of product prototype and mould design using CAD software. A tetrahedral finite element mesh of the product is exported as STL files. ML reads the files and creates models, and generates a scheme of the injection machine and mould design based on the CAD models. While all mould parts are discretized by triangular finite elements, a hybrid finite-element/finite-difference/control-volume numerical solution is used to simulate the mould filling. Then the data of simulation are sorted. The numerical results can be evaluated with FEA results visualiser; and melt flow during filling stage can be viewed. To simulate the true process, the user must correctly define the processing conditions. In order to optimise the moulding process parameters, the simulation results will be exported to the coupled optimisation system.

In this project, a mathematical and numerical model will be used to simulate the filling stage of injection moulding based on three-dimensional model. While this model is dealing with three-dimensional flow, 3D control volume method will be employed to track the flow front. To calculate the flow pattern, pressure and temperature profiles, the finite-element/finite-difference/control-volume method will be used to solve the momentum equation, continuity equation and energy equation.

3.5. Coupled Optimisation Model (COM)

This system couples two types of optimisation algorithms that include DOE, and SAPSO-based ANN, which have different natures in searching the optimal solution. The structure of COM can be seen in Figure.3.

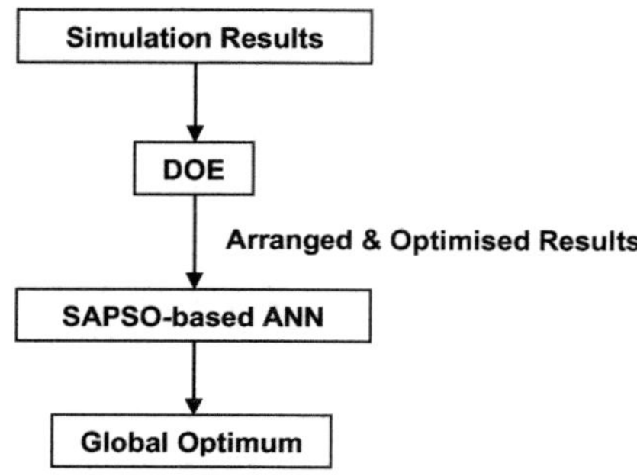

Figure.3. Flow diagram of COM

3.5.1. DOE

To identify the process condition and product components that influence product quality and productivity, DOE will be used in this study to find cause and effect relationship between the output and input experimental factors in a process. The DOE procedure includes four steps as following (Park & Ahn 2004):

- *Project definition:* identify the objective of the project and find the scope of the problem.

- *Screening:* reduction of the number of variables by identifying the key variables that affect product quality.

- *Optimization:* determination of the optimal values for various experimental factors.

- *Verification:* performing a follow-up experiment at the predicted best processing conditions to confirm the optimization results.

DOE approach can be divided into a full factorial design and a fractional factorial design.

- Full factorial design:

$$N_{full} = m(n_l)^k$$

- Fractional factorial design:

$$N_{fractional} = m(n_l)^{k-1}$$

In this study, k=5, n_l=2 ➔ N_{full}=2^5=32. In this five-factor experiment, there are 5 main effects (A,B,C,D, and E), 10 two-factor interactions (AB, AC, AD,…), 10 three-factor interactions (ABC, ABD, ADC,…), 5 four-factor interactions (ABCD, ACDE,…) and 1 five-factor interaction. However, higher – order interaction effects (interaction effects involving three or more factors) are very seldom significant. Therefore, the fractional experiment is designed to be able to consider main effects and two-way interactions. It is efficiently used in the screening DOE procedures when there are a large number of factors.

Alias Relationship for Fractional 2^5 Designs can be seen in Appendix B and 2^{5-1} Design in Appendix C (Yang & El-Haik 2003).

The results of DOE can be analyzed by evaluating the main effects for all factors and interactions for all of the two-way combinations. The main effect of the jth factor (Ej) and the interaction between the jth and kth factor Ijk) are calculated as follows (Park & Ahn 2004):

$$E_j = \frac{\sum_{i=1}^{N} l_{ij} R_i}{\sum_{i=1}^{N} R_i}$$

$$I_{jk} = \frac{\sum_{i=1}^{N} (l_{ij} l_{ik}) R_i}{\sum_{i=1}^{N} R_i}$$

Where N is the total number of experiments, Ri the response variable for the ith combination, lij equals -1 for the lower level, $+1$ to the upper level of the jth factor and is subjected to orthogonal condition as follows (Yang & El-Haik 2003):

$$\sum_{i=1}^{N} l_{ij} = 0 \quad (j=1,2,\ldots,N)$$

The results of the experimental designs are analysed by using analysis of variance (ANOVA). The ANOVA is utilised to investigate the relationship between a response variable and one or more independent variables. If the difference between the averages of the levels is greater than what could reasonably be expected from the variation that occurs within the level, it can be determined. The test for ANOVA generally uses a ratio of:

$$F = \frac{average(SS_{between})}{average(SS_{within})} = \left[\frac{1}{n_g - 1} \sum_{j=1}^{n_g} (n_j - 1) \right] \frac{SS_{between}}{SS_{within}}$$

where $SS_{between}$ and SS_{within} denote the square sum variation between groups, and the square sum variation within a group, respectively. Each term can be expressed as follows:

$$SS_{between} = n_j \sum_{j=1}^{n_g} (\overline{x_j} - \overline{x'})^2 \qquad , \qquad SS_{within} = \sum_{j=1}^{n_g} \sum_{i=1}^{n_i} (x_{ij} - \overline{x_j})^2$$

When applying the DOE, the best factor level settings and optimal output performance level are identified. On the other hand, to identify the optimal process condition, we can use Minitab software or Excel to analyse the DOE.

3.5.2. SAPSO-based ANN

At the latter phase of optimisation process, the data proceeded by DOE will be solved by the application of SAPSO-based ANN method. This is a hybrid solution initiated by the three original algorithms SA, PSO and ANN. Accordingly, all standard steps of a sample PSO algorithm, which is known as one of the most productive computational intelligence technique, presented by Da and Xiurun (2004) must be followed:

1. Initialise a number of particles with positions chosen randomly and velocities on d dimensions in a particular space.

2. For each particle, specify the fitness function to evaluate the desired optimisation of d variables:

$$fitness = -\sqrt{\frac{\sum_{i=1}^{np}\sum_{i=1}^{no}(y_i - o_{ij})^2}{nonp}}$$

The larger the fitness, the better set of weights and thresholds, which can be defiend as follow:

$$f\left(\sum_{i=1}^{n} w_i x_i - \theta\right) = 1/\left\{1 + \exp\left[-\left(\sum_{i=1}^{n} w_i x_i - \theta\right)\right]\right\}$$

3. Compare the finess evaluation of each particle with particle's p_{id}, if the current value is better than p_{id}, then set the value and location of p_{id} equal to those of the current value.

4. Compare the finess evaluation with the all population's p_{id}, if the current value is better than p_{id}, then set p_{id} to the array of index and value of the current particle.

5. Change the velocity and positon of the particle according to the two following equations:

$$v_i(t+1) = w v_i(t) + c_1 r_1(p_{id} - x_i(t)) + c_2 r_2(p_{gd} - x_i(t))$$

$$x_i(t+1) = x_i(t) + v_i(t+1)$$

6. return to step 2 and repeat until a pre – specified fitness function or a maximum number of iteration is met.

Concurrently, SA which is another intelligent algorithm, integrated into PSO method to explore the global optimum. Particularly, the flow of process is shown as follow (Da & Xiurun 2004):

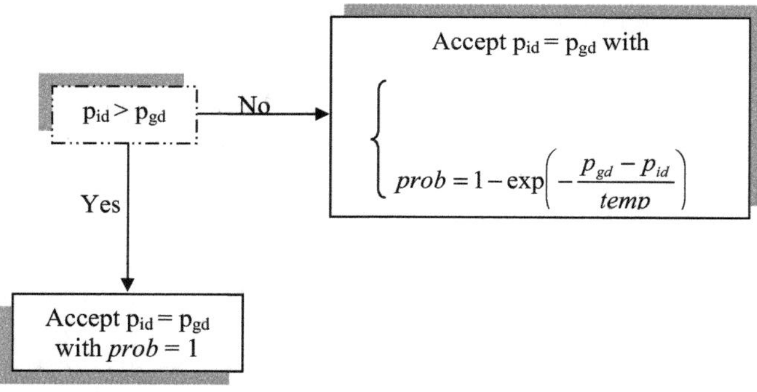

Figure.4. SAPSO logical sequence

The SAPSO method above is then employed to train a three – layer artificial neural network to generate a comprehensive SAPSO-based ANN method.

Overall, it takes approximately three minutes to complete a sample circle of SAPSO-based ANN experiment with the computer equipped a configuration as (Da & Xiurun 2004).

4. Requirements of the project

4.1. Timeframe

This research has been planned to carry out within 1 year by the nominated 4 students at the University of South Australia (UniSA). The detail timeframe for this study can be seen in Appendix D.

4.2. Supervision

The supervision of this project will be acquired from 2 supervisors; one is Dr. Ke Xing and another one is Prof. Lee Luong from the School of advanced Manufacturing and Mechanical Engineering, UniSA.

4.3. Facilities

To perform the study the required facilities will be provided by UniSA, including computer system installed with all required software, permission to have access to database for further literature and workshop during the experiment period, support for statistical analysis from School of Mathematics or Language and Teaching Unit (LTU) and so forth.

4.4. Estimated cost

The total cost for this project is estimated at $10,000 for 2 supervisors, trips to the industries and other expenses.

5. Significance of the project

This research aims to produce a comprehensive package for optimising the injection moulding process. Chen et al. (2009) claim that determining the optimal process setting is important because of potential influences on process efficiency, quality and the cost of products. The proposed approach couples computer simulations with different optimisation methods so it will be an effective tool to verify the optimal parameters for the entire process under multi-response considerations. Therefore, this study will propose a new efficient methodology with high flexibility for injection moulding to replace high-cost and time-consuming approaches. In addition, the use of this proposed integration system would provide detail information about the whole operation process in training.

6. Conclusion and future work

This research integrates the latest optimisation techniques: VR, DOE and SAPSO-based ANN to develop the most effective optimisation system for injection moulding process. As a result, the final outcome can be a fruitful comprehensive system applicable for a wide variety of products with high complexity 3D design in moulding industry. The experiment results are then compared with other approaches for official testification and self-evaluation before real-world application or manufacturing environment. However, as VR technique is still under development stage, this study may challenge with shortcomings and lack of technical support in building the system. Besides, the system is not included collision detection to find the possible defects of the mould design. In the fast-paced of industry development, there is no limit of improvement in any area. Therefore, it is necessary to do further researches that are more intensive to response the highly changing global market place. Future work in this study area should take into account cost estimation of moulding process and waste reduction to carry out the optimisation process in an economical pathway.

7. References

Bikas, A, Nikos Pantelelis, N & Kanarachos, A 2002, 'Computational tools for the optimal design of the injection moulding process', *Journal of Materials Processing Technology*, vol. 122, no. 1, pp. 112-126.

Blaxter, L, Hughes, C & Tight, M 2001, *How to research*, 2nd edn, Open University press, Philadelphia, USA.

Chang, RY & Yang, WH 2001, 'Numerical simulation of mold filling in injection molding using a three-dimensional finite volume approach', *International Journal for Numerical in Fluids*, vol. 37, pp. 125-148.

Chen, WC, Fu, GL, Tai, PH, Deng, WJ & Fan, YC 2007, 'ANN and GA-based process parameter optimisation for MIMO plastic injection moulding', *Proceedings of the 6th International Conference*, Machine Learning and Cybernetics, Hong Kong, pp. 1909-1917.

Chen, WC, Wang, MW, Fu, GL & Chen, CT 2008, 'Optimization of plastic injection molding process via Taguchi's parameter design method, BPNN, AND DFP', *Proceedings of the 7th International Conference,* Machine Learning and Cybernetics, Kunming, pp. 3315-3321.

Chen, WC, Fu, GL, Tai, PH & Deng, WJ 2009, 'Process parameter optimization for MIMO plastic injection molding via soft computing', *Expert Systems with Applications*, vol. 36, no. 2, pp. 1114-1122.

Chiang, KT & Chang, FP 2006, 'Application of grey-fuzzy logic on the optimal process design of an injection-molded part with a thin shell feature', *International Communications in Heat and Mass Transfer*, vol. 33, pp. 94-101.

Chiu, CC, Su, CT, Yang, GH, Huang, JS, Chen, SC, & Chen, NT 1997, 'Selection of optimal parameters in gas assisted injection molding using a neural network model and the Taguchi method', *International Journal of Quality Science*, vol. 2, no. 2, pp. 106-120.

Deng, WJ 2008, 'Integrating MRSN Ratio and ANP to Optimize Process Parameter of Multiple-response Injection Molding Process', *IEEE Service Operations and Logistics, and Informatics*, vol. 2, pp. 2736-2740.

Deng, YM, Zheng, D & Lu, XJ 2008, 'Injection moulding optimisation of multi-class design variables using a PSO algorithm', *International Journal of Advanced Manufacturing Technology*, vol. 39, no. 7-8, pp. 690-698.

Dong, C 2006, 'An equivalent medium method for the vacuum assisted resin transfer molding process simulation', *Journal of composite materials*, vol. 40, no. 13, pp.1193 – 1213.

Gerber, AG, Dubay, R & Healy, A 2006, 'CFD-based predictive control of melt temperature in plastic injection molding', *Applied Mathematical Modelling*, vol. 30, no. 9, pp. 884-903.

Kim, S, Turng, L 2004, 'Developments of three-dimensional computer-aided engineering simulation for injection moulding', *Modelling and simulation in materials science and engineering*, vol. 12, no. 3, pp. S151 – S173.

Kurtaran, H & Erzurumlu, T 2006, 'Efficient warpage optimization of thin shell plastic parts using response surface methodology and genetic algorithm', *The International Journal of Advanced Manufacturing Technology*, vol. 27, no. 5-6, pp. 468-472.

Lam, YC, Deng, YM & Au, CK 2006, 'A GA/gradient hybrid approach for injection moulding conditions optimization', *Engineering with Computers*, vol. 21, no. 3, pp. 193-202.

Lam, YC, Zhai, LY, Tai, K & Fok, SC 2004, 'An evolutionary approach for cooling system optimization in plastic injection moulding', *International Journal of Production Research*, vol. 42, no.10, pp. 2047-2061.

Lee KS & Lin JC 2006, 'Design of the runner and gating system parameters for a multi-cavity injection mould using FEM and neural network', *International Journal of Advanced Manufacturing Technology*, vol. 27, pp. 1089-1096.

Leedy, PD & Ormrod, JE 2005, *Practical research,* Pearson Merrill Prentice Hall, 8th edn, Columbus, Ohio, USA.

Liao, SJ, Chang, DY, Chen, HJ, Tsou, LS, Ho, JR, Yau, HT, Hsieh, WH, Wang, JT & Su, YC 2004, 'Optimal Process Conditions of Shrinkage and Warpage of Thin-Wall Parts', *Polymer Engineering and Science,* vol. 44, no. 5, pp. 917-928.

Llado', J & Sa'nchez, B 2008, 'Influence of injection parameters on the formation of blush in injection moulding of PVC', *Journal of Materials Processing Technology*, vol. 204, pp. 1-7.

Loera, VG, Castro, JM, Diaz, JM, Mondrago'n, OLC & Cabrera-R'ıos, M 2008, 'Setting the Processing Parameters in Injection Molding Through Multiple-Criteria Optimization: A Case Study', *IEEE Transactions on Systems, Man, and Cybernetics-Part C: Applications and Reviews,* vol. 38, no. 5, pp. 710-715.

Manzione, LT (ed.) 1987, *Applications of Computer Aided Engineering in Injection Moulding,* Macmillan, New York.

Ozcelik, B & Erzurumlu, T 2006, 'Comparison of the warpage optimization in the plastic injection molding using ANOVA, neural network model and genetic algorithm', *Journal of Materials Processing Technology,* vol. 171, pp. 437-445.

Oktem, H, Erzurumlu, T & Uzman, I 2007, 'Application of Taguchi optimization technique in determining plastic injection molding process parameters for a thin-shell part', *Material and Design*, vol. 28, no. 4, pp. 1271-1278.

Park, K and Ahn, JH 2004, 'Design of experiment considering two-way interactions and its application to injection molding processes with numerical analysis', *Journal of Materials Processing Technology*, vol. 146, no. 2, pp. 221-227.

Rosato, DV & DV Rosato 1995, *Injection Molding Handbook: The Complete Molding Operation Technology, Performance, Economics, 2nd edn,* Chapman & Hall, Massachusetts.

Sercer, M, Godec, D, Bujanic, B 2007, 'Application of Moldex3D for thin-wall injection moulding simulation', *AIP conference proceedings*, vol. 908, no. 1, pp. 1067 – 1072.

Shen, C, Wang, L & Li, Q 2007, 'Optimization of injection molding process parameters using combination of artificial neural network and genetic algorithm method', *Journal of Materials Processing Technology,* vol. 183, no. 2-3, pp. 412-418.

Shen C, Wang, L & Li Q 2007, 'Numerical Simulation of Compressible Flow with Phase Change of Filling Stage in Injection Molding', *Journal of Reinforced Plastics and Composite,* vol. 26, no. 4, pp. 353-372.

Shojaei, A 2006, 'Numerical simulation of three-dimensional flow and analysis of filling process in compression resin transfer moulding', *Composites Part A: Applied Science and Manufacturing,* vol. 37, pp. 1434-1450.

Su, CT & Chang, HH 2000, 'Optimization of parameter design: an intelligent approach using neural network and simulated annealing', *International Journal of Systems Science,* vol. 31, no. 12, pp. 1543-1549.

Subramanian, NR, Tingyu, L & Seng, YA 2005, 'Optimizing warpage analysis for an optical housing', *Mechatronics,* vol. 15, pp. 111-127.

Tseng, HY 2006, 'Welding parameters optimization for economic design using neural approximation and genetic algorithm', *International Journal Advanced Manufacturing Technology,* vol. 27, no. 9, pp. 897-901.

Turng, LS & Peic, M 2002, 'Computer aided process and design optimization for injection moulding', *Proceedings of the Institution of Mechanical Engineers, Part B: Journal of Engineering Manufacture,* vol. 216, no. 12, pp. 1523-1532.

Wu, CH & Liang, WJ 2005, 'Effects of geometry and injection-molding parameters on weld-line strength', *Polymer Engineering and Science,* vol. 45, no. 7, pp. 1021-1030.

Yang, K & El-Haik, B 2003, *Design for six sigma: a roadmap for product development,* McGraw-Hill, New York.

Zhai, M, Lam, YC & Au, CK 2006, 'Runner sizing and weld line positioning for plastics injection moulding with multiple gates', *Engineering with Computers,* vol. 21, pp. 218-224.

Zhou, H, Shi, S & Ma, B 2009, 'A virtual injection molding system based on numerical simulation, *International Journal Advanced Manufacturing Technology,* vol. 40, pp. 297-306.

Zhou, J & Turng, LS 2007, 'Process Optimization of Injection Molding Using an Adaptive Surrogate Model with Gaussian Process Approach', *Polymer Engineering and Science,* vol. 47, pp. 684-694.

Appendix A

Table of quality response comparison of different approach

	AMD	Standard Deviation				Mean	
	Length	Weight	Length	Warpage	Shrinkage	Warpage	Shrinkage
Other approach							
Proposed approach							
Improvement							

Appendix B

Alias Relationship for Fractional 2^5 Designs

I = ABCDE

↓

A = BCDE → AB = CDE   AC = BDE   AD = BCE   AE = BCD

↓

B = ACDE →   BC = ADE   BD = ACE   BE = ACD

↓

C = ABDE →   CD = ABE   CE = ABD

↓

D = ABCE →   DE = ABC

↓

E = ABCD

Appendix C

2^{5-1} Design

Run number	Factors														
	A	B	C	D	E	AB	AC	AD	AE	BC	BD	BE	CD	CE	DE	I=ABCDE
1	-1	-1	-1	-1	1	1	1	1	-1	1	1	-1	1	-1	-1	1
2	1	-1	-1	-1	-1	-1	-1	-1	-1	1	1	1	1	1	1	1
3	-1	1	-1	-1	-1	-1	1	1	1	-1	-1	-1	1	1	1	1
4	1	1	-1	-1	1	1	-1	-1	1	-1	-1	1	1	-1	-1	1
5	-1	-1	1	-1	-1	1	-1	1	1	-1	1	1	-1	-1	1	1
6	1	-1	1	-1	1	-1	1	-1	1	-1	1	-1	-1	1	-1	1
7	-1	1	1	-1	1	-1	-1	1	-1	1	-1	1	-1	1	-1	1
8	1	1	1	-1	-1	1	1	-1	-1	1	-1	-1	-1	-1	1	1
9	-1	-1	-1	1	-1	1	1	-1	1	1	-1	1	-1	1	-1	1
10	1	-1	-1	1	1	-1	-1	1	1	1	-1	-1	-1	-1	1	1
11	-1	1	-1	1	1	-1	1	-1	-1	-1	1	1	-1	-1	1	1
12	1	1	-1	1	-1	1	-1	1	-1	-1	1	-1	-1	1	-1	1
13	-1	-1	1	1	1	1	-1	-1	-1	-1	-1	-1	1	1	1	1
14	1	-1	1	1	-1	-1	1	1	-1	-1	-1	1	1	-1	-1	1
15	-1	1	1	1	-1	-1	-1	-1	1	1	1	-1	1	-1	-1	1
16	1	1	1	1	1	1	1	1	1	1	1	1	1	1	1	1

Appendix D

Timeframe of the project

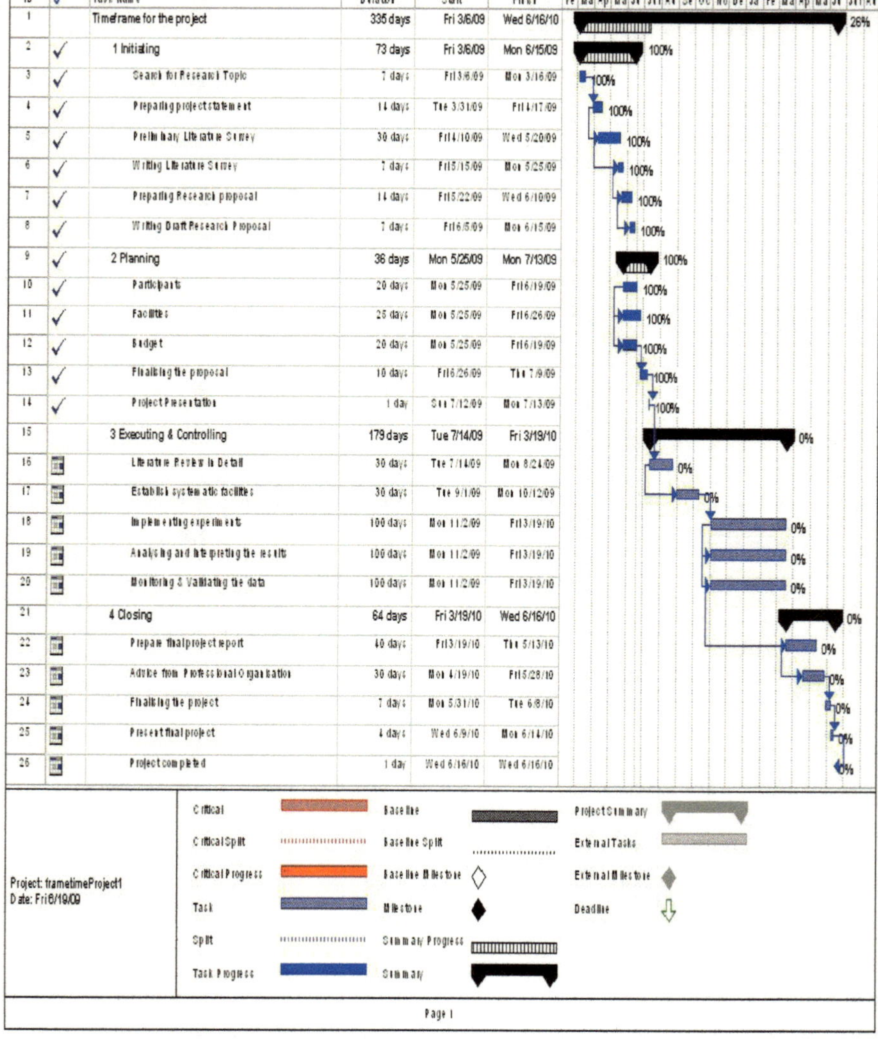

YOUR KNOWLEDGE HAS VALUE

- We will publish your bachelor's and
  master's thesis, essays and papers

- Your own eBook and book -
  sold worldwide in all relevant shops

- Earn money with each sale

Upload your text at www.GRIN.com
and publish for free